OXFORD
in association with

science
museum

time

OXFORD
UNIVERSITY PRESS

Great Clarendon Street, Oxford OX2 6DP

Oxford University Press is a department of the University of Oxford.
It furthers the University's objective of excellence in research, scholarship,
and education by publishing worldwide in

Oxford New York

Auckland Bangkok Buenos Aires Cape Town Chennai
Dar es Salaam Delhi Hong Kong Istanbul Karachi Kolkata
Kuala Lumpur Madrid Melbourne Mexico City Mumbai Nairobi
São Paulo Shanghai Taipei Tokyo Toronto

Oxford is a registered trade mark of Oxford University Press
in the UK and in certain other countries

British Library Cataloguing in Publication Data available

ISBN 0–19–910871-4 Hardback
ISBN 0–19–910872-2 Paperback

1 3 5 7 9 10 8 6 4 2

Printed in Italy

Acknowledgements

The publishers would like to thank:
For the Science Museum: David Rooney

All photos reproduced in kind permission of the Science and Society Picture Library
with the exception of the following:
Bridgeman Art Library; p8tr, 9
British Museum; p6bl
Digital Vision; p6tl, 20l, 21
English Heritage; p6-7c
Hemera Images; p8tr
National Maritime Museum; p18, 19cl & tl
The Ancient Art and Architecture Collection; p7tr, 22tr

OXFORD

in association with

science
museum

time

Contents

In the beginning

FOR OUR ANCIENT ancestors, knowledge of time was vital – they needed to find food or plant crops at exactly the right time of year. They judged the season by looking at the trees, and guessed the time of day from the Sun's position in the sky. But before long the Babylonians and others began to study the stars' regular movements and work out proper calendars.

↗ The seasons
Red autumn leaves told ancient peoples that there would be berries to gather, ground to clear, and that winter would soon be on its way.

↗ Sun god
This Babylonian relief shows people worshipping the Sun god Shamash. He has put an image of the Sun on an altar, while the real Sun, Moon and Venus shine above his head.

↗ God of learning

In ancient Egypt, calendars were produced by professional scribes. The god of scribes and learning was Thoth, an ibis-headed creature who carried a pen box.

← Stonehenge

Built and rebuilt during the period 3000–1600 BC, Stonehenge is Europe's most famous prehistoric monument. The stone circle was designed for rituals linked to the calendar. Its builders arranged the stones so that every Midsummer Day the Sun rises above an outlying stone called the Heel Stone, visible from the centre of the circle. With prehistoric technology, the builders must have worked very hard to produce this effect, and their work shows how important the calendar must have been to people in ancient Britain.

Calendar time

THE EARTH takes one solar year, or 365.24 days, to travel around the Sun. Calendars enable us to measure time by dividing up the year into days and months. Because the solar year does not last an exact whole number of days, modern calendars add an extra day in some years to keep the Sun and the calendar in step with each other.

➔ Julius Caesar
The Romans' first calendar drifted out of step with the Sun, so Caesar changed the calendar, bringing in a 365-day year, with an extra day every four years.

← Lunar calendar
The traditional Muslim calendar is based on the phases of the Moon. Lunar months last about 29.5 days, making a year of 354 days, so Muslim and western calendars do not correspond.

← Round and round
This 18th-century German perpetual calendar has rotating discs that provide all sorts of information – days and dates, feast days, signs of the Zodiac, and phases of the Moon.

← Book of Hours

In the Middle Ages, rich people used beautifully decorated prayer books called books of hours. A book of hours usually contained a calendar, so that people could find dates of saints' days and other religious festivals. This example is the month of March from a superb illustrated book of hours made for the Duke of Berri, France. At the top, the calendar includes the signs of the Zodiac (Pisces the two fish and Aries the ram) and the phases of the Moon. Below, peasants carry out the tasks needed at this time of year.

Sun and stars

IN THE NORTHERN hemisphere, the Sun seems to pass steadily across the sky from east to west. Thousands of years ago, people realized that they could measure time by setting up a pointer and watching its changing shadow, which fell towards the west in the morning and towards the east in the afternoon. They had made the first simple sundial.

↑ Shadow clock
The ancient Egyptians produced shadow clocks like this. They lined up the baseboard in an east–west direction and read off the time by looking at the position of the shadow cast by the thin crosspiece.

HORAS NON NUMERO NISI SERENAS

VII
VIII
IX
X
XI XII I II
III
IIII
V
VI

1650

← On the wall
For hundreds of years, many churches, town halls and other important buildings had sundials. Any passer-by could look at the shadow cast by the gnomon (pointer) and read the time of day. This sundial is showing a time of 2.30 in the afternoon. The Latin inscription at the top means 'I count only the hours that are serene (peaceful)'.

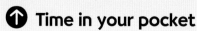

↑ Time in your pocket
Portable sundials were also popular. This one from China is made of wood. The gnomon folds away to make it easy to carry, and it has a built-in compass.

⬊ In the dark

A sundial was useless at night, but you could tell the time with an instrument called a nocturnal. To use the nocturnal, you viewed the Pole Star through the central hole while lining up the pointer with a pair of stars in the Great Bear constellation. You could then read the time on one of the instrument's scales. This stunning gold-plated nocturnal was probably made in the late 17th century and is decorated with dragon-like serpents, perhaps to remind the user of another constellation, Draco, the dragon.

Water, fire and sand

ANYTHING THAT MOVES at a steady, known rate can be used to measure time. The people of ancient Egypt, Babylon and China used the flow of water. The Europeans developed the sand glass, like the modern egg-timer, in which powdered rock or sand trickles from one part of a glass container to another.

← Water clock
Water dripped out of this Egyptian container through a hole in the base. A scale showed how much water was left and how much time had passed.

→ Candle clock
One way to measure time is to work out how long a candle takes to burn. The length of the candle can then be divided into one-hour sections.

← Round and round
This model is based on a design by the 11th-century Chinese writer Su Sung. At its heart is a water clock that drives a series of gears. These power a model of the stars and planets, showing how they move through the sky.

⬆ By the hour

From the Middle Ages until the 18th century, sand glasses were one of the most popular ways of measuring time. They could be used to measure anything from the time it took to cook a meal to the length of a sermon. This set of glasses was designed to be used together. When the first glass is empty, half an hour has passed; the second glass empties after one hour, and so on.

MECHANICAL CLOCKS had appeared by the 13th century. They measured time using a mechanism of gear wheels powered by steadily falling weights. Often they had no faces, but were connected to bells to strike the hour. Then clockmakers connected the wheels to pointers or hands, so that people could read the time on a dial.

← Multi-purpose clock

Early clockmakers were fascinated by complex dials, which gave the user all sorts of information. This example is the face of a large clock at Hampton Court Palace, London. As well as a 24-hour dial to tell the time, it displays the current sign of the zodiac, the month and date, and the phase of the Moon. Elaborate clocks like this were expensive to make and were usually only seen in important houses and churches.

↑ Turret timepiece

This large 17th-century clock was designed to be installed in a tower. Its face has only an hour hand – the clock is not accurate enough to measure minutes.

↓ Wooden wonder

Some craftworkers liked to make clocks almost completely out of wood. In this Swiss clock of 1669 the wheels, pivots, framework and beautifully painted case are all wooden.

↗ Night piece

Some clocks, like this clever 16th-century timepiece, were designed for use at night. As well as an alarm, there are raised studs so that you can 'feel' the time in the dark.

Swing time

THE ITALIAN SCIENTIST Galileo noticed when watching a pendulum that it always took the same amount of time to swing. He realized that you could make an accurate clock by controlling its mechanism with a swinging pendulum. In 1656 Dutchman Christiaan Huygens designed the first successful pendulum-controlled clock. The next year, when the clock was built, it was accurate to within about five minutes a day.

● Rocking steadily

This mechanism is based on a 17th-century drawing by Galileo. The swinging pendulum on the right is connected at its upper end to a curved metal strip. This strip engages with a series of pins on the upper wheel, called the escape wheel. The escape wheel is linked to the clock's other wheels. As the pendulum swings, the metal strip moves, releasing the escape wheel for a moment before swinging back to 'catch' another of the pins. This means that the escape wheel moves at a regular rate.

⬅ ⬆ Spring power

Complex clocks like this show the skill of the 17th-century clockmaker. The dials show information such as phases of the Moon. The mechanism, shown above, is powered by springs, contained in metal barrels. The barrels are connected to little chains, which wind around cone-shaped drums called fusees.

➡ Weight-driven

Many clocks, like this Japanese example from about 1700, were driven by hanging weights. The user had to move the weights upwards at regular intervals to keep the clock going.

Time at sea

TELLING THE TIME accurately on a sea journey is important because it helps you to work out your position precisely. In the 18th century, Parliament put up a prize for the person who could come up with a way of working out longitude at sea. The search for an accurate sea-clock began.

➡ Harrison's first chronometer

In the hope of winning the prize, British craftsman John Harrison spent 40 years making clocks, or chronometers, that would keep good time at sea. He made his first sea-clock in the 1730s. Its four dials indicate the hours, minutes, seconds, and days of the month. It was taken on a trial voyage to Lisbon in Portugal in 1736. It did well, but Harrison insisted that he could build an even more accurate timekeeper.

⬇ Third effort

Harrison's second timekeeper did not satisfy him either, so he spent 19 years working on a third. This chronometer was also highly accurate, and it contained one invention, caged roller bearings to reduce friction, similar to bearings still used in machines today.

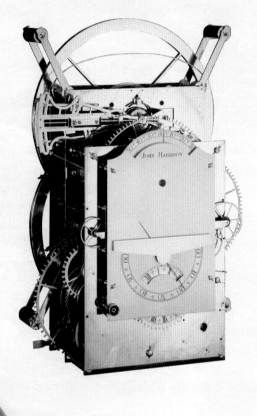

⬆ 'The Watch'

Harrison's fourth timekeeper was 'the Watch', much smaller than his earlier sea-clocks and even more accurate. Its mechanism uses tiny jewels, such as diamonds and rubies, to reduce friction.

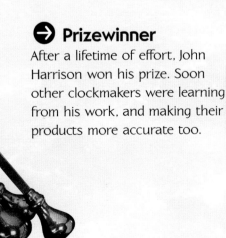

➡ Prizewinner

After a lifetime of effort, John Harrison won his prize. Soon other clockmakers were learning from his work, and making their products more accurate too.

Super accuracy

FROM THE FINE hand-made timepieces of the 19th century to today's digital watches, timekeepers have become more and more accurate. In the home, clocks and watches that contain a crystal of the mineral quartz are highly accurate. But the most precise of all are atomic clocks, which are controlled by vibrating atoms of caesium and which can be accurate to within one second in 3 million years.

← 'Big Ben'

The clock in the Houses of Parliament in Westminster, London, has several innovations to make it keep good time, including a special mechanism to stop the wind slowing the hands down. There is also a very simple method of making the clock run slower or faster – placing coins on the pendulum or taking them off.

↑ Quartz watch

At the heart of a modern wrist watch is a tiny crystal of quartz that vibrates steadily, keeping each timepiece super-accurate.

← Radio control

Some modern clocks and watches can receive radio signals from an atomic clock. The signal allows the clock to be adjusted if it has lost or gained time.

← Finishing line

At major sports events, a high-speed camera is used to photograph the competitors as they cross the finishing line. This allows officials to see who has won even when the race is very close. In addition, the photograph is linked to a quartz stopwatch. This can be accurate to within hundredths of a second, and can give a precise, split-second timing for every athlete who finishes the race.

Glossary

Babylonian Person from Babylon, an ancient city in Mesopotamia (part of modern Iraq) which was the centre of an advanced civilization from 3000 BC onwards

chronometer Instrument for measuring time accurately; such as the accurate sea-clocks made to aid navigation in the 18th, 19th and 20th centuries

constellation Several stars that form a group in the night sky when viewed from the Earth

day The time taken for the Earth to make one complete turn on its axis, 24 hours

dial The face of a clock or watch

friction A resisting force produced when one object moves against another

gear Wheel with notches, teeth, or pins around the edge, used to transmit movement in a machine such as a clock

gnomon Pointer on a sundial

longitude Angular distance east or west of a line, known as the Prime Meridian, which passes through the North and South Poles and also through Greenwich, England; longitude provides a way of describing the east-west position of any object on the Earth's surface

lunar month The time taken for the Moon to travel once around the Earth, 29.53 days

nocturnal Instrument for measuring the time at night by making sightings of the stars; the word nocturnal is also used to mean 'by night'

pendulum A freely swinging weight on the end of a rod or thread; the regular swing of a pendulum is used to control the movement of many clocks

perpetual calendar Calendar that can be adjusted so that it can be used in any year

quartz Common rock-forming mineral, occurring in crystals; when an electrical current is applied to a quartz crystal, the crystal vibrates at a fixed rate and can be used to control a watch or clock accurately

ritual Ceremony, often carried out as part of a religion

sand glass A glass instrument used to measure time; a sand glass is normally made up of two linked sections containing a fixed amount of sand, which runs steadily from one section to the other

solar year The time taken for the Earth to travel around the Sun, 365.24 days

vibrate To move backwards and forwards, such as shaking or swinging

water clock Instrument that measures time by means of a steady flow of water

Index

Page numbers in *italic* type refer to illustrations.